SOLOMON GRUNDY,
BORN ON ONEDAY

A FINITE ARITHMETIC PUZZLE

Poor old Solomon Grundy. Born on Monday and died on Sunday, just seven days later. Poor *young* Solomon Grundy. He had a short life. Or did he? This delightful Young Math Book explores the possibilities of Solomon Grundy's life and treats it as a puzzle in finite arithmetic—in this case, an arithmetic with only seven numbers in it (one for each day of the week).

Children are encouraged to work out the possibilities for Mr. Grundy's life by constructing a week-clock with seven numerals on it. In the end, they discover they can imagine a very long life indeed for Solomon Grundy. In the process they experience both adding and subtracting in a finite arithmetic system, and are encouraged to locate other examples of finite arithmetic in their lives.

Tomie de Paola has contributed wonderfully humorous and informative illustrations for this inviting additon to the Young Math series.

PRESENTING

SOLOMON GRUNDY,
BORN ON ONEDAY

A Finite Arithmetic Puzzle

by
Malcolm E. Weiss

illustrated by
Tomie de Paola

Thomas Y. Crowell Company · New York

YOUNG MATH BOOKS

Edited by Dr. Max Beberman, Director of the Committee on School Mathematics Projects, University of Illinois

BIGGER AND SMALLER

CIRCLES

COMPUTERS

THE ELLIPSE

ESTIMATION

FRACTIONS ARE PARTS OF THINGS

GRAPH GAMES

LINES, SEGMENTS, POLYGONS

LONG, SHORT, HIGH, LOW, THIN, WIDE

MATHEMATICAL GAMES FOR ONE OR TWO

ODDS AND EVENS

PROBABILITY

RIGHT ANGLES: Paper-Folding Geometry

RUBBER BANDS, BASEBALLS AND DOUGHNUTS: A Book About Topology

STRAIGHT LINES, PARALLEL LINES, PERPENDICULAR LINES

WEIGHING & BALANCING

WHAT IS SYMMETRY?

Edited by Dorothy Bloomfield, Mathematics Specialist, Bank Street College of Education

ANGLES ARE EASY AS PIE

AREA

AVERAGES

BASE FIVE

BINARY NUMBERS

BUILDING TABLES ON TABLES: A Book About Multiplication

EXPLORING TRIANGLES: Paper-Folding Geometry

A GAME OF FUNCTIONS

HOW DID NUMBERS BEGIN?

HOW LITTLE AND HOW MUCH: A Book About Scales

LESS THAN NOTHING IS REALLY SOMETHING

MAPS, TRACKS, AND THE BRIDGES OF KÖNIGSBERG: A Book About Networks

MEASURE WITH METRIC

NUMBER IDEAS THROUGH PICTURES

ROMAN NUMERALS

SHADOW GEOMETRY

666 JELLYBEANS! ALL THAT? An Introduction to Algebra

SOLOMON GRUNDY, BORN ON ONEDAY: A Finite Arithmetic Puzzle

SPIRALS

STATISTICS

3D, 2D, 1D

VENN DIAGRAMS

YES-NO; STOP-GO: Some Patterns in Logic

Published simultaneously in Canada by Fitzhenry & Whiteside Limited, Toronto. Manufactured in the United States of America.

Library of Congress Cataloging in Publication Data Weiss, Malcolm E. Solomon Grundy, born on Oneday. SUMMARY: Discusses a solution to an arithmetic problem based on the rhyme about Solomon Grundy's life. 1. Modular arithmetic—Juv. lit. [1. Arithmetic. 2. Time] I. De Paola, Thomas Anthony, illus. II. Title.
QA247.35.W43 513′.6 76-26560 ISBN 0-690-01275-6

1 2 3 4 5 6 7 8 9 10

SOLOMON GRUNDY,
BORN ON ONEDAY

Solomon Grundy,

Born on Monday,

Christened on Tuesday,

Married on Wednesday,

Took sick on Thursday,

Worse on Friday

Died on Saturday,

Buried on Sunday;

This is the end of Solomon Grundy.

So goes a Mother Goose nursery rhyme.

It's a rhyme with puzzles in it.

Poor old Solomon Grundy! He lived a short life, didn't he? Born on Monday. Married two days later. Sick the day after that. Died two days after that. Lived his whole life in seven days. Poor *young* Solomon Grundy!

Is that the story of Solomon Grundy? Do you think it could be true? Did he get married when he was two days old? What did he say to his bride? "Goo, goo"? or "Gurgle, gurgle"? Where's the catch?

And what about christening? "Christened on Tuesday," says the rhyme. That means that Solomon Grundy was given his name on Tuesday.

Well, that seems to make sense. A baby *would* be named soon after it was born. But christening means more than just parents choosing a name for their baby. Christening is a special naming ceremony. It takes place on baby's first trip to church, usually when he or she is about three weeks old. Yet the rhyme tells us Solomon Grundy was christened on Tuesday, one day after he was born.

HE WAS BORN YESTERDAY.

Or does it? Let's take a careful look at what the rhyme really tells us—and what it *doesn't* tell us.

The rhyme says Solomon Grundy was christened on Tuesday. What Tuesday?

There is one Tuesday in each week. A week is seven days. If you count seven days from Tuesday, you get back to Tuesday again. If you count a week from any day, you get back to the same day again.

So Solomon Grundy might have been christened on the Tuesday one week and a day, or two weeks and a day, or three weeks and a day after he was born. He might have been christened any number of weeks and a day after he was born. All the rhyme tells us is that it was on *some* Tuesday that Solomon Grundy was christened. And there are lots of Tuesdays.

How can we keep track? How can we show that when we count a week from any day, we come back to the same day of the week again?

Well, we could just write "Monday, Tuesday, Wednesday, Thursday, Friday, Saturday, Sunday, Monday, Tuesday. . . ." out on a line. But there are a few things wrong with that idea. We'd get tired of writing. Even if we didn't get tired, we'd run out of paper.

What's more, this idea doesn't show the pattern of the days of the week—the way they keep repeating and coming back to the beginning every seven days. And since there are only seven different names for the days, it's a waste of time to keep writing them over and over again.

SUNDAY	MONDAY	TUESDAY	WEDNESDAY	THURSDAY	FRIDAY	SATURDAY
	1	2	3	4	5	6
7	8	9	10	11	12	13
14	15	16	17	18	19	20
21	22	23	24	25	26	27
28	29	30				

What about writing the days of the week out just once at the top of a page? Then we could number the days underneath, the way a calendar does.

That does show a pattern of weeks. The weeks are stacked, one on top of another. All the Mondays are in one line, all the Tuesdays are in another line, and so on. But there is an even better way to show that the first day of the new week comes right after the last day of the old week.

The days of the week seem to go around and around in a circle, like the numbers on a clock. We can put the days of the week in a circle. We can make a sort of "week-clock" to help keep track of the days of the week. All we need is a paper plate, a cardboard hand for the clock, and a paper fastener.

E NEED A PENCIL, TOO.

Poor young Solomon Grundy could have used a week-clock like this one. Since Solomon Grundy's week began with Monday, the day he was born, we'll start our clock with Monday.

An ordinary clock uses numbers to count the hours. Solomon Grundy's week-clock counts the days of the week. We can use numbers to stand for these days. Monday is Oneday, or 1 for short. Tuesday is Twoday, or 2 for short, Wednesday is Threeday, and so on right around to Sunday = Sevenday = 7.

There are only seven different days of the week. So the week-clock needs only seven different numbers. When you count the week-clock way, you say "1, 2, 3, 4, 5, 6, 7" and then you come back to 1 again. The biggest number is 7. Then the numbers start all over again.

When you count the way we usually count, you say "1, 2, 3, 4, 5, 6, 7, 8, 9, 10, 11, 12, 13, 14. . . ." and on and on. No matter how long you count, you never run out of numbers. Our usual counting system has no end. It has no biggest number. Each number is one bigger than the number before it.

THERE'S NO END TO IT!

The main difference between the week-clock counting system and our usual counting system is that the week-clock does have a biggest number. The week-clock system is called a **finite** number system. Finite means that it has just so many numbers and no more. When we reach 7 we are finished with the numbers of the week-clock counting system.

Let's see if we can find some ways week-clock numbers and ordinary numbers are like each other, and some ways that they are different from each other.

One way of putting together numbers is by adding them. In our usual counting system $4 + 1 = 5$. We get the answer by starting with 4 and moving ahead one step in our counting system.

MOVE ONE STEP.
4
5

What's 4 + 1 in the week-clock system? We start with the week-clock hand pointing at 4. To add 1, we move it ahead one step. Does the answer look familiar?

Now try 5 + 2. In our usual system, of course, that is 7. What is it on the week-clock?

So far, week-clock arithmetic gives the same answers as ordinary arithmetic. Now, how about 7 + 1? Are the answers the same in both arithmetics? Be careful! Use the paper-plate clock to help you. Remember to start with the clock hand at 7, and move one step ahead.

That is right! In week-clock arithmetic $7 + 1 = 1$!

What happens when you try $7 + 2$ in both arithmetics? $7 + 3$?

Why does week-clock arithmetic give such strange answers? When the week-clock hand goes a step past seven, it has made a complete turn on the clock. That means it gets back to Oneday again. Monday is always one day after Sunday, Tuesday is always two days after Sunday, and so on.

Finite arithmetic is arithmetic-in-a-circle.

To help you compare ordinary arithmetic and week-clock arithmetic, make some charts like these on your own piece of paper.

	1	2	3	4	5	6	7
1	2	3	4	5	6	7	8
2	3	4	5	6	7	8	9
3	4	5	6	7	8	9	10
4	5	6	7	8	9	10	11
5	6	7	8	9	10	11	12
6	7	8	9	10	11	12	13
7	8	9	10	11	12	13	14

	1	2	3	4	5	6	7
1	2	3	4	5	6	7	1
2	3	4	5	6	7	1	2
3	4	5	6	7	1	2	3
4	5	6	7	1	2	3	4
5	6	7	1	2	3	4	5
6	7	1	2	3	4	5	6
7	1	2	3	4	5	6	7

ORDINARY	WEEK-CLOCK
1 + 2 = 3	1 + 2 = 3
2 + 1 = 3	2 + 1 = 3
3 + 2 = 5	3 + 2 = 5
2 + 3 = 5	2 + 3 = 5

Sometimes the answers are the same, and sometimes they are different.

ORDINARY	WEEK-CLOCK
5 + 4 = 9	5 + 4 = 2
4 + 5 = 9	4 + 5 = 2
2 + 6 = 8	2 + 6 = 1
6 + 2 = 8	6 + 2 = 1
5 + 6 = 11	5 + 6 = 4
6 + 5 = 11	6 + 5 = 4

This is how the week-clock works. And it's how the puzzle of Solömon Grundy works. "Christened on Tuesday" does not mean Solomon Grundy was a day old when he was christened. It could mean Tuesday after three complete turns of the week-clock. Each turn stands for a week.

"Married on Wednesday" could mean the Wednesday twenty years after the Monday Solomon was born. That would be about a thousand turns on the week-clock.

The rhyme really doesn't tell us much about how long Solomon Grundy lived. We've just added twenty years to his life! We can spin the week-clock and add lots more. That may help solve the puzzle. Was

Solomon really sick one day and worse the next? Or could he have been sick once on a Thursday, and more seriously sick on a Friday many years later?

You can add other facts to Solomon Grundy's life.
There are so many things the rhyme doesn't tell us.

ATE NINE WATERMELONS ON WEDNESDAY,
AND WAS HUNGRY ON THURSDAY.

1813
SU
M
TU
W
TH
F
SA
7
1
2
3
4
5
6

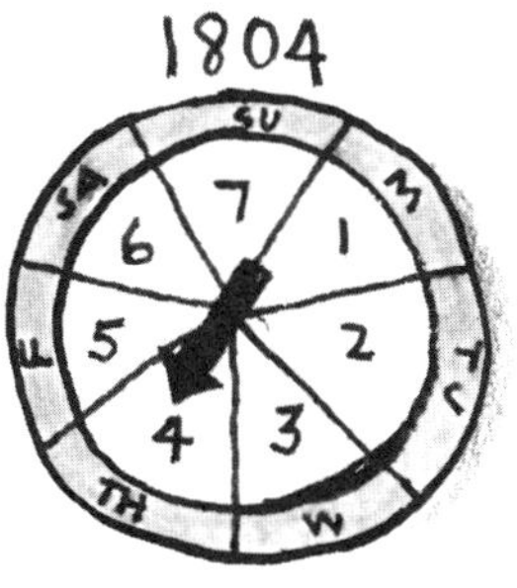
1804
SU
M
TU
W
TH
F
SA
7
1
2
3
4
5
6

You can find out more about finite numbers.
Are there other ways they work the same as ordinary numbers? There was a clue on page 25.

NEXT LET'S TRY ADDING ZERO TO NUMBERS IN BOTH SYSTEMS.
ORDINARY
1+2 = 2+1
3+2 = 2+3
5+4 = 4+5
2+6 = 6+2
5+6 = 6+5
WEEK-CLOCK
1+2 = 2+1
3+2 = 2+3
5+4 = 4+5
2+6 = 6+2
5+6 = 6+5

Sometimes the answers are different, but the two systems work the same.

Can you think of other finite number systems? How many numbers do you need to work them?

FOUR HOURS UNTIL SCHOOL'S OUT.
12 1 2 3 4 5 6 7 8 9 10 11
11+4=3
IN FOUR MORE SEASONS IT WILL BE WINTER AGAIN.
WINTER - 1
SPRING - 2
SUMMER - 3
FALL - 4
1+4=1

About the author

Malcolm Weiss is the author of numerous books and magazine articles for children on science. He has also written *666 Jellybeans: All That? An Introduction to Algebra,* in the Young Math series.

Mr. Weiss lives in Whitefield, Maine, with his wife, and daughters Margot and Rebecca.

About the illustrator

Tomie de Paola received his Bachelor of Fine Arts degree from Pratt Institute and also studied at the Skowhegan School of Painting and Sculpture, Skowhegan, Maine, under the noted painter Ben Shahn. He holds advanced degrees from the California College of Arts and Crafts and Lone Mountain College in San Francisco.

In addition to illustrating and writing children's books, Mr. de Paola's interests include the theater, cooking, teaching, traveling, his old house and his two cats. He has had many one-man shows of his prize-winning paintings and drawings.

Tomie de Paola now lives in New Hampshire where he is an associate professor of visual arts at New England College, in Henniker.